MY Learning Stations

TEACHER GUIDE
DIFFERENTIATED MATH INSTRUCTION

Grade 4

 **Education**

Bothell, WA • Chicago, IL • Columbus, OH • New York, NY

Education

Send all inquiries to:
McGraw-Hill Education
8787 Orion Place
Columbus, OH 43240

ISBN: 978-0-02-117181-1
MHID: 0-02-117181-5

Printed in the United States of America.

3 4 5 6 7 8 9 QDB 17 16 15 14 13 12

Common Core State Standards© Copyright 2010. National Governors Association Center for Best Practices and Council of Chief State School Officers. All rights reserved.

Our mission is to provide educational resources that enable students to become the problem solvers of the 21st century and inspire them to explore careers within Science, Technology, Engineering. and Mathematics (STEM) related fields.

The *McGraw·Hill* Companies

Contents

i

Components of
MY Learning Stations

Differentiated Learning Stations

My Learning Stations is a collection of activity cards, literature, games, graphic novels, and problem-solving cards that can be used with each chapter of the *My Math* student edition. Differentiated instruction strategies for each component are provided within this teacher guide.

 The activities in *My Learning Stations* align to the Common Core State Standards and support the Standards for Mathematical Practices.

Carrying Case

Stores all of the components that make up the learning stations kit.

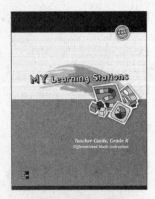

Learning Station Guide

Includes instructions for using each component with approaching-level, on-level, and beyond-level students. Also includes answers and blackline masters.

KEY: Approaching Level   On Level   Beyond Level

Games

Reinforce, challenge, and extend mathematical concepts being taught in the Student Edition. English is on one side and Spanish is on the other. Some of the materials needed to play the games are included in the kit.

* Provided in kit

 Graphic Novel

Introduce and revisit real-world mathematical situations. Animations bring the math content to life.

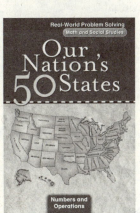

 Real-World Problem Solving Readers (RWPS)

Fiction and non-fiction readers extend problem-solving skills and strategies and make real-world connections. Three leveled readers—approaching level, on level, and beyond level—are included in the kit.

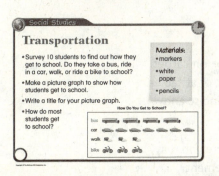

 Activity Cards

Give students an opportunity to learn mathematics through cross-curricular connections. English is on one side and Spanish is on the other.

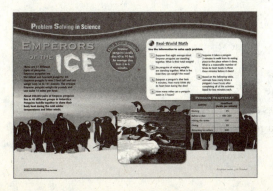

 Problem Solving

Tie problem solving to other disciplines with real-world applications. English is on one side and Spanish is on the other. (Grades 3–5 only)

Managing
MY Learning Stations

What are math learning stations?

Math learning stations are set-up as places in the classroom where students work independently or in small groups exploring, practicing, or extending mathematical concepts. When working in math learning stations, students are engaged in problem-solving activities; they are reasoning; representing their thinking; communicating; and making connections between mathematical practices and content. *My Learning Stations* uses games, real-world problem solving readers, activity cards, and graphic novels to meet individual student needs. While students are involved in learning stations, the teacher works with individuals or meets with small groups in order to further differentiate math instruction.

How do I organize my classroom?

Evaluate the space in your classroom and think about how it can be arranged to best promote independent and small group instruction.

- Place the learning station easel in an area that is easily accessible to students.

- Make sure all supplies needed for each activity are available at the learning station easel.

- Ensure that the space you have chosen allows for easy cleanup.

- Encourage students to complete the activity at their desks, on the floor, on carpet tiles, at the computer, at an interactive whiteboard, at a bulletin board, or at tables. Allow students to spread out around the room to manage the level of noise and movement.

- Be sure the places where students choose to work are visible from where you may be working with a small group to allow you to monitor student engagement and progress.

What might this look like?

While you are working with a small group of eight students on the day's math lesson, two students are on the floor playing a game, four students are watching a graphic novel, and four students are at desks reading a real-world problem-solving reader.

How do I get started?

During the first few weeks of school, develop and model procedures for using math learning stations.

- Discuss with the class what happens during math learning station time. Write student responses on a chart that you can display in the classroom. For example:

Math Learning Station Time

See	Hear	Think
students talking about math	math vocabulary being used	This is fun!
students taking turns	students explaining what they did	I can do math!
students sharing materials	students asking questions	I know how to do that!

- Math learning station time comes after a whole-class math lesson. This is not a time to introduce new concepts. It is a time for students to further explore, practice, or extend a previously taught math concept. To ensure students will be able to work independently during math learning station time, model what you expect students to do at each math learning station.

- For younger students and English language learners, create a brief list with simple drawings or pictures to help them remember what to do at each learning station.

- The time spent at a learning station will vary with the age of the students and the time of the school year. Kindergarten students may spend about 15 minutes in each learning station while first through fifth grade students may spend about 20 minutes.

- Going through all the learning stations in a chapter may take several days or weeks depending on the needs of the students in your class.

- Create a system for managing any blackline masters needed at the learning stations.

  - Label a folder for each activity and put the blackline masters needed for that activity in the folder.

- Create a system for managing learning station work that is completed and ready for review. For example:

  - Label folders for each student or activity.

  - Label a box *Finished* in which students place completed work.

  - Use student logs or journals for recording.

  - Bring students back together in a large group and have them share any work product that was created at the learning station or tell what they did at the learning station.

- Decide on a rotation chart format to direct who goes to which learning station. If students are going to two learning stations, use a bell or timer to signal when to change learning stations.

What might a rotation chart look like?

Following is an example of one type of rotation chart you might use.

- For grades K-2, divide a piece of chart paper into five sections. Label the sections with the following titles: "Meet with Me"; Game; Graphic Novel; Real-World Problem Solving Reader; and Activity Card.

- For grades 3-5, divide a piece of chart paper into six sections. Label the sections with the following titles: "Meet with Me"; Game; Graphic Novel; Real-World Problem Solving Reader; Activity Card; and Problem-Solving Card.

- Write each student's name on a note card.

- Place student name cards in each learning station section on the chart to indicate which learning station they will be doing.

- If students are going to two learning stations, you will need to make a different color name card for each student and number the name cards 1 and 2. Then place the student name cards in each learning station section on the chart to indicate the order in which students do the stations.

How do I differentiate the learning stations?

Math learning stations allow you to differentiate for individual- and small-group needs of the students in the classroom.

- Determine station groupings

  - Use informal assessments (including observations) to determine individual student needs.

  - Decide whether groups will be homogeneous or heterogeneous.

  - Remember to keep reassessing and regrouping throughout the year.

  - Determine whether students will be working in a group or individually.

  - Use the buddy system to help with individualized work.

- After assessing individual student needs, refer to specific activity strategies in this Teacher Guide for differentiated instruction suggestions.

  AL identifies strategies for students who have needs that are below or approaching grade level.

  OL identifies strategies for students on grade level.

  BL identifies strategies for students who have needs that are above or beyond grade level.

- As students progress in concept understanding, they may repeat activities using different strategies, approaches, tools, and models.

> **What might this look like?**
> Strategies for playing a game may include having students who are *approaching level* work in teams of two to play, students who are *on level* play the game with the rules as written on the game board, and students who are *beyond level* bump the game up by using higher numbers.

Activities in MY Learning Stations

Activities in My Learning Stations (continued)

The following learning station activities provide differentiated instructional strategies for helping students identify place value of whole numbers to millions, compare and order, and round whole numbers.

 4.NBT.2 Use after Lesson 1-2

Game Greater Number Game

Materials 40 index cards, pencil, place-value chart

AL Have students play the game with the rules as written. Allow them to use a place-value chart to help them compare numbers.

OL Have students play the game with five-digit numbers.

BL Have students play the game with six-digit numbers.

Extend the Game Have students play the game using numbers in the millions. Differentiate the game by following the directions above.

 4.NBT.2 Use after Lesson 1-3

Graphic Novel Wet Weekend Fun!

Materials blackline master pages 31 and 32, pencil

At the beginning of the chapter, watch the graphic novel as a class and discuss the following questions.

- What are Morgan, Kendra, and Carlos trying to figure out? *which water park is more popular*

- If you were in Morgan's position, how would you decide which park to go to? *Sample answer: I would choose the park with the most water slides.*

AL Have students solve the graphic novel problem using blackline master page 31.

OL Have students solve the graphic novel problem using blackline master page 31. On the back of the blackline master, have students extend the graphic novel by adding another problem that uses math from the chapter. Have other students solve the new problem.

BL Have students solve the graphic novel problem using blackline master page 31. Have students write their own graphic novel that uses math from the chapter using blackline master page 32. Have other students solve the new graphic novel.

KEY: Approaching Level 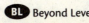 On Level **BL** Beyond Level * Provided in kit

Real-World Problem Solving Reader *Rivers and Mountains of the United States*

Summary *Rivers and Mountains of the United States* compares geographic features of rivers and mountains throughout the United States. The book also discusses effects of rivers and mountains on population changes. Students will use number sense and place value skills to answer questions.

Materials paper, pencil

AL Have students read the story and with a partner answer Exercises 1, and 3–6.

OL Have students read the story and answer Exercises 1–6.

BL Have students read the story and answer Exercises 1–6. Have students find the lengths of the Missouri River, Mississippi River, Rio Grande River, Colorado River, Rocky Mountains, Appalachian Mountains, and Sierra Mountains. Then order the lengths from *least* to *greatest*.

Activity Card Tell Me How

Materials paper, pencil

AL Have students complete the activity as written.

OL Have students complete the activity using six-digit numbers.

BL Have students complete the activity using seven-digit numbers.

Problem Solving Card Creatures Under the Sea

Materials index card, paper, pencil

Activate Prior Knowledge Before your students begin this activity, as a class discuss rivers.

- Name some marine mammals that you know. Which ones have fur? *whales, seals; seals*

- Name some characteristics that all mammals share. *Sample answers: They are warm-blooded; they give birth to live young.*

AL Have students read the card and with a partner answer Exercises 1–6.

OL Have students read the card and answer Exercises 1–6. On an index card, write the following problems for the students to solve: Which dolphin had the greatest population?; If you round the population of the Northern Fur Seal to the nearest ten thousand, how many are there?

BL Have students read the card and answer Exercises 1–6. Then have students create a word problem that uses math from the chapter and information found in the text and picture. Have other students solve the new problem.

CHAPTER 2 Add and Subtract Whole Numbers

The following learning station activities provide differentiated instructional strategies for helping students solve multi-digit addition and subtraction problems.

 4.NBT.4 Use after Lesson 2-7

Game Make a Big Difference

Materials paper, pencil, blank spinner* (labeled 0–9), blackline master page 33

AL Have students make a game sheet subtracting three-digit numbers.

OL Have students play the game with the rules as written.

BL Have students make a game sheet subtracting six-digit numbers.

Extend the Game Have students play the game using sums and the lowest sum wins. Differentiate the game by following the directions above.

 4.NBT.4 Use after Lesson 2-7

Real-World Problem Solving Reader *The Olympic Games*

Summary *The Olympic Games* describes the history of the modern Games, shows maps with major cities in which the modern Games have occurred, and presents a variety of facts about the Games. Students will use addition and subtraction to solve problems.

Materials paper, pencil

AL Have students read the story and with a partner answer Exercises 1–4.

OL Have students read the story and answer Exercises 1–4.

BL Have students read the story and answer Exercises 1–5.

CCSS 4.NBT.5 Use after Lesson 2-9

Real-World Problem Solving Reader *Oceans: Into the Deep*

Summary *Oceans: Into the Deep* introduces students to the various zones of the ocean, including the distance that each zone is from the water's surface and what lives in each zone. Graphs show animals in each zone, and a comparison of the depths of the world's oceans. Students will use subtraction to solve problems.

Materials paper, pencil

AL Have students read the story and with a partner answer Exercises 3 and 6.

OL Have students read the story and answer Exercises 1, 3, and 6.

BL Have students read the story and answer Exercises 1–6.

Copyright © The McGraw-Hill Companies, Inc.

3

4.NBT.4   Use after Lesson 2-7

Activity Card Climb Every Mountain

Materials index cards, markers, paper, pencil, base-ten blocks

AL While completing the activity, allow students to use base-ten blocks to help them add and subtract.

OL Have students complete the activity as written.

BL Have students complete the activity as written. Then have students round each mountain height to the nearest thousand and complete the activity again.

 4.NBT.4   Use after Lesson 2-9

Problem-Solving Card Ready, Set, Click!

Materials paper, pencil

Activate Prior Knowledge Before your students begin this activity, as a class discuss photography.

- What type of cameras have you or people you know used? *Sample answers: digital, flash, disposable, cell phone*

- Before the digital process, how were photographs developed? *using chemicals*

AL Have students read the card and with a partner answer Exercises 1–7.

OL Have students read the card and answer Exercises 1–7. On an index card, write the following problems for students to solve: How long has it been since the start of the digital revolution?; Suppose you buy one of each type of camera with two bills. How much change would you receive?

BL Have students read the card and answer Exercises 1–7. Then have students figure out the greatest number of cameras they can buy for $50 without any change.

The following learning station activities provide differentiated instructional strategies for helping students understand the relationship between multiplication and division.

 4.OA.4 Use after Lesson 3-7

Game Factor Power

Materials 20 index cards, crayons, 1-inch graph paper for each player, multiplication table

AL While playing the game, allow students to use a multiplication fact table to help them find factors.

OL Have students play the game with the rules as written.

BL While playing the game, have students write a multiplication sentence to match each array. Both must be correct in order to gain 1 point.

Extend the Game Have students play the game laying arrays end to end to get to a finish line. Differentiate the game by following the directions above.

 4.NBT.5 Use after Lesson 3-6

Graphic Novel Dog Walking Dollars

Materials blackline master pages 34 and 35, pencil

At the beginning of the chapter, watch the graphic novel as a class and discuss the following questions.

· What are Teresa and Ethan trying to figure out? *if Teresa will have enough money to buy the DVD before volleyball practice starts*

· Have you ever tried to earn money to buy something you want? What kind of jobs did you do? *Sample answers: babysitting, mowing the lawn, washing the dog*

· How much did you earn? *Sample answers: $5, $10*

AL Have students solve the graphic novel problem using blackline master page 34.

OL Have students solve the graphic novel problem using blackline master page 34. Have students write their own graphic novel that uses math from the chapter using blackline master page 35. Have other students solve the new graphic novel.

BL Have students solve the graphic novel problem using blackline master page 34. On the back of the blackline master, have students extend the graphic novel by adding another problem that uses math from the chapter. Have other students solve the new problem.

CCSS 4.NBT.5   Use after Lesson 3-6

Real-World Problem Solving Reader *Class Project*

Summary *Class Project* focuses on surveys taken by students in six regions of the country. These students compare the results of their surveys. In the process, students will read maps and charts and interpret data using multiplication and division.

Materials paper, pencil

AL Have students read the story and with a partner answer Exercises 3–4.

OL Have students read the story and answer Exercises 2–4.

BL Have students read the story and answer Exercises 1–5.

CCSS 4.OA.4   Use after Lesson 3-7

Activity Card Factor Sifter

Materials markers, paper, multiplication table

AL While completing the activity, allow students to use a multiplication table to help them find factors.

OL Have students complete the activity as written.

BL Have students complete the activity as written. Then have students complete the activity again for numbers 51–100.

CCSS 4.NBT.5   Use after Lesson 3-3

Problem-Solving Card Pop Culture

Materials paper, pencil

Activate Prior Knowledge Before your students being this activity, as a class discuss soda pop.

- What makes soda pop fizzy? *carbonated water*

- Who started adding flavors to soda water? *pharmacists*

AL Have students read the card and with a partner answer Exercises 1–7.

OL Have students read the card and answer Exercises 1–7. On an index card, write the following problems for students to solve: If you bought 3 cases of soda in 1894, how many bottles would you have bought?; How many "home-packs" would a case of soda bottles make?

BL Have students read the card and answer Exercises 1–7. Have students determine how many ounces are in a 12-pack and in a case of 24 12-ounce soda cans.

The following learning station activities provide differentiated instructional strategies for helping students multiply up to four digits by a one-digit number.

 CCSS 4.NBT.5 Use after Lesson 4-9

Game High and Low

Materials blank number cube* (labeled 1–6), blackline master page 36, pencil

AL While playing the game, allow students to multiply a two-digit number by a one-digit number.

OL Have students play the game with the rules as written.

BL While playing the game, have students multiply a four-digit number by a one-digit number.

Extend the Game Have students multiply their products by 10. Differentiate the game by following the directions above.

 CCSS 4.NBT.5 Use after Lesson 4-9

Graphic Novel Musical Math

Materials blackline master pages 37 and 38, pencil

At the beginning of the chapter, watch the graphic novel as a class and discuss the following questions.

- Have you ever loaded music onto a MP3 player or other portable device? *Answers will vary.*

- Were you able to fit every song that you wanted onto the device? Explain. *Sample answer: No, I had more songs that I wanted than I had space for on my MP3 player.*

AL Have students solve the graphic novel problem using blackline master page 37.

OL **BL** Have students solve the graphic novel problem using blackline master page 37. Then have students work together to write a graphic novel that uses math from the chapter using blackline master page 38. Have the students act out their graphic novel to the class, having the class solve the problem.

 CCSS 4.NBT.5 Use after Lesson 4-11

Graphic Novel A Pressing Problem

Materials blackline master pages 39 and 40, pencil

At the beginning of the chapter, watch the graphic novel as a class and discuss the following questions.

- How many days does Kendra say it takes to mix and cook the jelly beans? *12 days*

- How does Alyssa know that is the same as 3 weeks? *There are 7 days in a week; 21 ÷ 7 = 3 weeks.*

AL Have students solve the graphic novel problem using blackline master page 39.

OL Have students solve the graphic novel problem using blackline master page 39. On the back of the blackline master, have students extend the graphic novel by adding another problem that uses math from the chapter. Have other students solve the new problem.

BL Have students solve the graphic novel problem using blackline master page 39. Have students write their own graphic novel that uses math from the chapter using blackline master page 40. Have other students solve the new graphic novel.

4.NBT.5   Use after Lesson 4-11

Activity Card Low or High Roller?

Materials 0–5 number cube*, paper, pencil

AL Have students complete the activity creating a three-digit number.

OL Have students complete the activity as written.

BL Have students complete the activity creating a six-digit number.

4.NBT.5   Use after Lesson 4-9

Problem-Solving Card Stamp Collecting

Materials paper, pencil

Activate Prior Knowledge Before your students begin this activity, as a class discuss stamps.

- What kinds of pictures have you seen on stamps? *Sample answers: flags, flowers, baseball players, birds*

- What values of stamps have you seen? *Sample answers: 42 cents, 39 cents*

AL Have students read the card and with a partner answer Exercises 1–4.

OL Have students read the card and answer Exercises 1–5, and 7.

BL Have students read the card and answer Exercises 1–7. Then have students create a word problem that uses math from the chapter and the current stamp value. Have other students solve the new problem.

4.NBT.5   Use after Lesson 4-11

Problem-Solving Card Emperors of the Ice

Materials paper, pencil

Activate Prior Knowledge Before your students begin this activity, as a class discuss penguins.

- What types of penguins can you name? *Sample answer: Emperor penguins*

- Where do Emperor penguins usually live? *In cold environments; Antarctica*

AL Have students read the card and with a partner answer Exercises 1, 3, 4, and 6.

OL Have students read the card and answer Exercises 1–6. Then have students create a word problem that uses math from the chapter and information found in the text and chart. Have other students solve the new problem.

BL Have students read the card and answer Exercises 1–6. Then have students figure out the range of the number of heartbeats a penguin could have during an average dive, using the 3-to-6 minute time range.

The following learning station activities provide differentiated instructional strategies for helping students multiply a two-digit number by a two-digit number.

CCSS 4.NBT.5   Use after Lesson 5-4

Graphic Novel Basketball Brainteaser

Materials blackline master pages 41 and 42, pencil

At the beginning of the chapter, watch the graphic novel as a class and discuss the following questions.

- How many points do you earn for making a basket in basketball? How many points do you make for a free-throw in basketball? *2 points; 1 point per free-throw*

- What is Carmen trying to figure out? *The number of points she scored during the basketball season.*

AL Have students solve the graphic novel problem using blackline master page 41.

OL Have students solve the graphic novel problem using blackline master page 41. Have students write their own graphic novel that uses math from the chapter using blackline master page 42. Have other students solve the new graphic novel.

BL Have students solve the graphic novel problem using blackline master page 41. On the back of the blackline master, have students extend the graphic novel by adding another problem that uses math from the chapter. Have other students solve the new problem.

 4.NBT.5   Use after Lesson 5-5

Real-World Problem Solving Reader *Expanding the United States*

Summary *Expanding the United States* reviews the settlement of the West. It includes details of major historical events, travel routes, distances, and types of transportation. Students will use multiplication to solve problems.

Materials paper, pencil

AL Have students read the story and with a partner answer Exercises 1–3, 5, and 6.

OL Have students read the story and with answer Exercises 1–3, 5, and 6.

BL Have students read the story and answer Exercises 1–6.

 4.NBT.5   Use after Lesson 5-6

Real-World Problem Solving Reader *What is Recycling?*

Summary *What is Recycling?* focuses on reducing, reusing, and recycling trash. Students will interpret charts and graphs and use multiplication skills to answer questions and draw conclusions.

Materials paper, pencil

AL Have students read the story and with a partner answer Exercises 1 and 3.

OL Have students read the story and answer Exercises 1–4.

BL Have students read the story and answer Exercises 1–5.

CCSS 4.NBT.5 Use after Lesson 5-4

Activity Card Field Trip

Materials paper, pencil

AL While completing the activity, allow students to draw rectangular arrays and/or area models to help them multiply two two-digit numbers.

OL Have students complete the activity as written.

BL Have students complete the activity with the trip cost per person being a three-digit number.

The following learning station activities provide differentiated instructional strategies for helping students find whole numbers quotients with up to four-digit dividends and one-digit divisors, including remainders.

 4.NBT.6   Use after Lesson 6-6

Game Mission: Division

Materials blank spinner* (labeled 0–9), base-ten blocks, index cards, crayons or markers, pencil

AL While playing the game, allow students to use base-ten blocks to help them divide.

OL Have students play the game with the rules as written.

BL Have students write their remainders as fractions.

Extend the Game Have students make game sheets with three-digit dividends and one- or two-digit divisors. Differentiate the game by following the directions above.

 4.NBT.6   Use after Lesson 6-10

Activity Card Long Song Contest

Materials blank number cube* (labeled with any number between 0–9), paper, pencil

AL Have students complete the activity as written.

OL Have students roll the number cube four times and make a four-digit number, then complete the activity as written.

BL Have students roll the number cube five times and make a five-digit number, then complete the activity as written.

 4.NBT.6   Use after Lesson 6-11

Problem-Solving Card A Desert Safari

Materials paper, pencil

Activate Prior Knowledge Before your students being this activity, as a class discuss deserts.

• What do all deserts have in common? *Sample answer: They are hot and dry.*

• What are some deserts animals that are on the endangered species? *Sample answer: elephants*

AL Have students read the card with a partner and answer Exercises 1–6.

OL Have students read the card and answer Exercises 1–7. Then have students create a word problem that uses math from the chapter and information found in the text and chart. Have other students solve the new problem.

BL Have students read the card and answer Exercises 1–7. Then have students make a chart showing how much one of each animal would eat per day, and per week. Round all remainders to the nearest ten.

CHAPTER 7 Patterns and Sequences

The following learning station activities provide differentiated instructional strategies for helping students understand how patterns are used in mathematics.

CCSS 4.OA.5 Use after Lesson 7-2

Graphic Novel A Successful Season

Materials blackline master pages 43 and 44, pencil

At the beginning of the chapter, watch the graphic novel as a class and discuss the following questions.

- What is the basketball team trying to figure out? *Sample answer: How many points they will score in tomorrow's game.*

AL Have students solve the graphic novel using blackline master page 43.

OL BL Have students solve the graphic novel problem using blackline master page 43. Then have students work together to write a graphic novel that uses math from the chapter to solve using blackline master page 44. Have the students act out their graphic novel to the class, having the class solve the problem.

CCSS 4..OA.5 Use after Lesson 7-8

Activity Card Exercise Equations

Materials index cards, pencil

AL Have students complete the activity as written.

OL Have students write a word problem using addition and a word problem using multiplication.

BL Have students write a word problem with two operations.

CCSS 4.OA.5 Use after Lesson 7-7

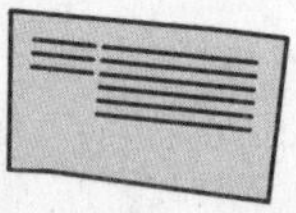

Problem-Solving Card A Visit to the Supermarket

Materials paper, pencil

Activate Prior Knowledge Before your students begin this activity, as a class discuss the supermarket.

- What kinds of foods do supermarkets sell? Which of those foods are healthy? *Sample answers: chips, fruit, soda, vegetables; fruit and vegetables*
- How are the foods organized? *Sample answer: Fruits and vegetables are together. Canned foods are on shelves. All of the frozen foods are together.*

AL Have students read the card and with a partner answer Exercises 2, 4, and 5.

OL Have students read the card and answer Exercises 1, 2, 4, 5, and 6.

BL Have students read the card and answer Exercises 1–6.

CHAPTER 8 Fractions

The following learning station activities provide differentiated instructional strategies for helping students find equivalent fractions, fractions in simplest form, mixed numbers, and improper fractions.

 CCSS 4.NF.2   Use after Lesson 8-9

Game Mixed Number Match

Materials blackline master pages 45–54, fraction circles or tiles, scissors

AL Have students use fraction circles or fraction tiles to model each fraction if needed.

OL Have students play the game with the rules as written.

BL Have students use the mixed number cards from the game and make equivalent mixed numbers of each one.

Extend the Game Have students come up with another equivalent mixed number in order to get a point. Differentiate the game by following the directions above.

 CCSS 4.NF.2   Use after Lesson 8-7

Real-World Problem Solving Reader *Life in the United States*

Summary *Life in the United States* focuses on population and the four regions established by the United States Census Bureau. These regions should not be confused with geographical regions. Students will use fractions to answer questions and represent data.

Materials paper, pencil

AL Have students read the story and with a partner answer Exercises 1–4.

OL Have students read the story and answer Exercises 1–4.

BL Have students read the story and answer Exercises 1–6.

The following learning station activities provide differentiated instructional strategies for helping students add and subtract like fractions and mixed numbers and multiply fractions by whole numbers.

 4.NF.3c   Use after Lesson 9-7

Activity Card   Voting Age

Materials   list of names and birthdays of students in the class, poster board, markers, paper pencil

AL   Have students complete the activity with a partner.

OL   Have students complete the activity as written. Have students display the ages of everyone in the classroom from least to greatest on the poster board.

BL   Have students complete the activity as written. Then create a bar graph showing the ages and display the bar graph on the poster board.

 4.NF.3d   Use after Lesson 9-7

Activity Card   I Will Take Two, Please

Materials   recipes from cookbooks or cooking magazines, paper, pencil

AL   Have students complete the activity with a partner.

OL   Have students complete the activity as written.

BL   Have students complete the activity as written. Then have students take their doubled recipe and double it again using multiplication instead of addition.

The following learning station activities provide differentiated instructional strategies for helping students understand the relationship between fractions and decimals.

CCSS 4.NBT.6   Use after Lesson 10-6

Game Fractoes and Decimoes

Materials blackline master pages 55–56, scissors, index cards

AL On the index cards, make hint cards that shows the decimal, word, and pictorial form of each fraction from $\frac{1}{10}$ to $\frac{10}{10}$. Allow students to use the hint cards while playing the game.

OL Have students play the game with the rules as written.

BL Have students play the game using only the decimals and fractions tiles, rather than pictorial tiles.

Extend the Game Have students make tiles for hundredths fractions. Make sure that for each hundredths fraction they make there is an equivalent word form tile and an equivalent decimal form tile. Differentiate the game by following the directions above.

CCSS 4.NF.7   Use after Lesson 10-7

Activity Card Olympic Race

Materials index cards, blank number cube* (labeled with numbers between 0 and 9), pencil

AL Have students determine their scores to the tenths place.

OL Have students complete the activity as written.

BL After students have completed the activity as written, have them reshuffle the cards and race to place the scores in order from least to greatest.

CCSS 4.NF.5   Use after Lesson 10-6

Problem-Solving Card Decimal Note-ation

Materials paper, pencil

Activate Prior Knowledge Before your students begin this activity, as a class discuss musical notes.

• What are musical notes based on? *fractions*

• Which musical note is equivalent to 1? *a whole note*

AL Have students read the card and with a partner answer Exercises 1–4.

OL Have students read the card and answer Exercises 1–4. Then have students create a word problem that uses math from the chapter and information found in the text and chart. Have other students solve the new problem.

BL Have students read the card and answer Exercises 1–9. Explain to students that the eighth note comes after the quarter note and have them write the eighth note as an equivalent fraction and decimal.

CHAPTER 11 Customary Measurement

The following learning station activities provide differentiated instructional strategies for helping students solve problems involving customary measurement and customary conversions.

 4.MD.1   Use after Lesson 11-9

Real-World Problem Solving Reader *Strange but True*

Summary *Strange but True* focuses on unusual creatures in a variety of animal groups. Students will use mathematical operations and measurement conversions to help solve problems.

Materials paper, pencil

AL Have students read the story and with a partner answer Exercises 2, 3, and 5.

OL Have students read the story and answer Exercises 2, 3–5.

BL Have students read the story and answer Exercises 1–6.

 4.MD.2   Use after Lesson 11-9

Real-World Problem Solving Reader *Ancient Giants of the Forest*

Summary *Ancient Giants of the Forest* focuses on large, old trees, particularly giant sequoias. Students will use measurement to solve problems.

Materials paper, pencil

AL Have students read the story and with a partner answer Exercises 1 and 2.

OL Have students read the story and answer Exercises 1, 2, and 4.

BL Have students read the story and answer Exercises 1–5.

 4.MD.1   Use after Lesson 11-6

Problem-Solving Card Tide Pool Ecosystems

Materials paper, pencil

Activate Prior Knowledge Before your students begin this activity, as a class discuss tide pools and creatures found in an aquarium.

• Name some creatures you know that live in tide pools. *Sample answers: starfish, crabs, mussels*

• Name a sea plant that appears in tide pools. *Sample answer: kelp*

AL Have students read the card and with a partner answer Exercises 1–7.

OL Have students read the card and answer Exercises 1–7. Have students determine how many ounces of gravel are needed for the 30-gallon and 40-gallon tanks.

BL Have students read the card and answer Exercises 1–7. Then have students create a word problem that uses math from the chapter and information found in the text and picture. Have other students solve the new problem.

16

CHAPTER 12

Metric Measurement

The following learning station activities provide differentiated instructional strategies for helping students solve problems involving metric measurement and metric conversions.

 4.MD.1   Use after Lesson 12-6

Activity Card Gas Conservation

Materials two blank number cubes* (labeled with numbers between 0 and 9), paper, pencil

AL Have students complete the activity playing only 5 rounds.

OL Have students complete the activity as written.

BL Have students complete the activity as written. Then have students take their total number of kilometers traveled and determine how many hours their trip would be at 90 km/h.

 4.MD.1   Use after Lesson 12-6

Problem-Solving Card Coral Reefs

Materials paper, pencil

Activate Prior Knowledge Before your students begin this activity, as a class discuss coral reefs.

• Where are coral reefs? *in the ocean*

• Who relies on the coral reef for food and shelter? *plants, fish*

AL Have students read the card and with a partner answer Exercises 1–7.

OL Have students read the card and answer Exercises 1–7. Then have students create a word problem that uses math from the chapter and information found in the text and picture. Have other students solve the new problem.

BL Have students read the card and answer Exercises 1–7. Have students conduct research to find the length in meters of three more animals that live in coral reefs and convert their lengths to centimeters and millimeters.

CHAPTER 13 Perimeter and Area

The following learning station activities provide differentiated instructional strategies for helping students determine perimeter and area.

 4.MD.3 Use after Lesson 13-4

Game Area Guess

Materials ruler with both inches and centimeters, pencil, blackline master page 57

AL Have students measure the length of the figures first and estimate using the length to guide their answers.

OL Have students play the game with the rules as written.

BL Have students use figures that are combinations of rectangles to estimate and measure.

Extend the Game Have students play the game using inches instead of centimeters and measure to the nearest inch. Differentiate the game by following the directions above.

 4.MD.3 Use after Lesson 13-5

Graphic Novel Having a Class Pet

Materials blackline master pages 58 and 59, pencil

At the beginning of the chapter, watch the graphic novel as a class and discuss the following questions.

- What is the class trying to figure out? *which aquarium to buy for the class pet*

- Why is it important to buy the right size aquarium for the class pet, the guinea pig?
  Sample answers: If you get one too small, the guinea pig could not move around.

AL Have students solve the graphic novel problem using blackline master page 58.

OL Have students solve the graphic novel problem using blackline master page 58. On the back of the blackline master, have students extend the graphic novel by adding another problem that uses math from the chapter. Have other students solve the new problem.

BL Have students solve the graphic novel problem using blackline master page 58. Have students write their own graphic novel that uses math from the chapter using blackline master page 59. Have other students solve the new graphic novel.

Copyright © The McGraw-Hill Companies, Inc.

18

CCSS 4.MD.3   Use after Lesson 13-5

Problem-Solving Card Walls with History

Materials index card, paper, pencil

Activate Prior Knowledge Before your students begin this activity, as a class discuss forts.

- Why do people build forts? *for shelter and protection*

- What famous forts have you heard of? *Sample answer: Fort Knox*

AL Have students read the card and with a partner answer Exercises 1, 3–7.

OL Have students read the card and answer Exercises 1, 3–7. On an index card, write the following problem for the students to solve: What is the total distance around Fort Clatsop?; What is the area of Fort Clatsop?

BL Have students read the card and answer Exercises 1–7.

The following learning station activities provide differentiated instructional strategies for helping students solve problems involving symmetry.

 4.G.3   Use after Lesson 14-10

Game Reflections and Symmetry

Materials 30 index cards, mirror*, paper, pencil

AL Have students work in pairs to help each other draw symmetrical shapes.

OL Have students play the game with the rules as written.

BL Have students make cards with shapes that are all asymmetrical.

Extend the Game Have students play the game using midlines instead of mirrors. Differentiate the game by following the directions above.

 4.G.3   Use after Lesson 14-10

Activity Card Butterfly Symmetry

Materials white paper, markers, scissors, mirror*

AL Have students complete the activity as written.

OL Have students complete the activity as written. Then have students draw a face following the same directions.

BL Have students complete the activity as written. Then have students draw a face but instead of using a mirror have them use a midline.

 4.G.3   Use after Lesson 14-10

Problem-Solving Card Symmetry in Nature

Materials paper, pencil

Activate Prior Knowledge Before your students begin this activity, as a class discuss types of symmetry.

• What items around the room have line(s) of symmetry? *Sample answers: desk, chair, window*

AL Have students read the card and with a partner answer Exercises 1–5.

OL Have students read the card and answer Exercises 1–5. Then have students create a word problem that uses math from the chapter and information found in the text and pictures. Have other students solve the new problem.

BL Have students read the card and answer Exercises 1–5. Have students look for other pictures of nature that show symmetry then explain how they know.

Answers

Answers and sample answers are provided for all the Graphic Novel, Real World Problem-Solving Reader, Activity Card, and Problem-Solving Card differentiated strategies.

CHAPTER 1 Place Value

 Graphic Novel Wet Weekend Fun!

Slide Island; Sample answer: She wants to go to the more popular park. Slide Island had 89,868 visitors last year which is greater than Wave City's visitors last year, 79,416.

 Check student's word problem for accuracy and reasonableness.

 Check student's graphic novel for accuracy and reasonableness.

 Real-World Problem Solving Reader *Rivers and Mountains of the United States*

1. 161 climbing walls; 122 + 39 = 161
2. The friend is incorrect because 21,120 > 4,500; Check student's drawing for reasonableness.
3. Lake Onalaska
4. Standard form: 1,885 miles; word form: one thousand, eight hundred eighty-five miles; expanded from: 1,000 + 800 + 80 + 5 miles
5. Rocky Mountain Range; The length of the Rocky Mountain Range is 3,000 miles. The length of the Appalachian Mountains is 1,600 miles. 3,000 > 1,600.
6. 7,000 feet

BL least to greatest: Sierra Mountains 400 miles, Colorado River 1450 miles, Appalachian Mountains 1,600, Rio Grande River 1,885 miles, Missouri River 2,350 miles, Mississippi River 2,565 miles; Rocky Mountains 3,000 miles

 Activity Card Passing Notes in Class

Check student's instructions on how to compare numbers using a number line for accuracy and reasonableness.

 Problem-Solving Card Creatures Under the Sea

1. Northern fur seal; 900,000 + 80,000 + 8,000; nine hundred eighty-eight thousand
2. no; There are more gray whales since 20,869 > 20,000.
3. more; 1,500 pounds more
4. Seal, 5,314 feet; dolphin, 1,000 feet; sea lion, 400 feet
5. Yes; 131,826 rounded to the nearest thousand is 132,000.
6. California sea lion; gray whale; Hawaiian monk seal; spinner dolphin; and spotted dolphin

 Spotted dolphin, 990,000

 Check student's word problem for accuracy and reasonableness.

CHAPTER 2 Add and Subtract Whole Numbers

Real-World Problem Solving Reader *The Olympic Games*

1. $776 - 580 = 196$ years
2. silver medals; $19 + 6 + 5 = 30$, $18 + 2 + 3 = 23$, $30 > 23$
3. The United States scored more points in the third and fourth quarters. $17 + 29 < 52 + 74$
4. $2,191 - 216 - 1,975$ medalss
5. about half $\left(\dfrac{2191}{3869}\right)$; US medals over total medals

Real-World Problem Solving Reader *Oceans: Into the Deep*

1. 175 grams of salt
2. Yes; Sample answer: 5,280 can be rounded to 5,000. So, 2 miles is about 10,000 feet. The *Titanic* lies about 13,000 feet below the surface of the water.
3. $13,100 - 3,300 = 9,800$ feet deep
4. $1,700 - 1,200$
5. Tan's claim is not reasonable. Since there are about 5,000 feet in a mile, there are about 25,000 feet in 5 miles. The Mariana Trench is more than 36,000 feet deep. That is more than 5 miles deep.
6. $35,000 - 18,000 = 17,000$ feet

Activity Card Climb Every Mountain

Check student's work to see if they added and subtracted correctly.

Problem-Solving Card Ready, Set, Click!

1. Sample answer: 2 digital, 1 underwater, and 1 flash
2. $2 \times \$10 + 3 \times \$4 + 3 \times \$6$; $50
3. $20
4. outdoor and black-and-white
5. $15

6. black-and-white and outdoor
7. Sample answer: Emily can buy 3 digital cameras or she can buy 6 black-and-white cameras.

 OL Sample answer: 30 years, $2011 - 1981 = 30$; Sample answer: 2 twenty-dollar bills, $60

BL Check student's word problem for accuracy and reasonableness.

CHAPTER 3 Understand Multiplication and Division

Graphic Novel Dog Walking Dollars

Yes; Each week Teresa earns $3 \times \$5$, or $15. In two weeks, she will earn $30. $30 > \$25$

 OL Check student's word problem for accuracy and reasonableness.

BL Check student's graphic novel for accuracy and reasonableness.

Real-World Problem Solving Reader *Class Project*

1. Six of the 50 states took the survey. $50 - 6 = 44.$ $\frac{44}{50}$ states did not take the survey.

2. surfing; basketball

3. 27 trophies; multiplication

4. football; $4 \times 4 = 16$

5. 64 hours each week; Sample answer: Emily practices for 2 hours four times a week, 7 people took the survey and add in Emily. $2 \times 4 \times 7 = 56$, $56 + 8 = 64 =$ hours each week

Activity Card Factor Sifter

2, 3, 5, 7, 11, 13, 19, 23, 29; Sample answer: These numbers are not multiples of 2, 3, and 5.

BL 53, 59, 61, 71, 73, 77, 79, 83, 89, 91; Sample answer: These numbers are not multiples of 2, 3, and 5.

Problem-Solving Card Pop Culture

1. 45¢

2. 315¢

3. 15

4. $24

5. 12: 2 home-packs

6. 5

7. 4 ways

OL 72 bottles; 4

BL 144 ounces; 288 ounces

Graphic Novel Musical Math

2,132 minutes; $467 \times 4 = 1{,}868$; $4{,}000 - 1{,}868 = 2{,}132$

OL Check student's word problem for accuracy and reasonableness.

BL Check student's graphic novel for accuracy and reasonableness.

Graphic Novel A Pressing Problem

1. 6,300 jelly beans

2. 12,600 jelly beans

OL Check student's word problem for accuracy and reasonableness.

BL Check student's graphic novel for accuracy and reasonableness.

Activity Card Low or High Roller?

Check student's paragraphs about estimations for reasonableness.

 Problem-Solving Card Stamp Collecting

1. 87¢
2. 2 stamps
3. 65¢
4. 96¢

5. 390 stamps
6. Sample answers: two 24¢ and one 37¢ stamps; or one 24¢, one 37¢, and six 4¢ stamps
7. 100¢ or $1

BL Check student's word problem for accuracy and reasonableness.

 Problem-Solving Card Emperors of the Ice

1. 528 lb
2. 252 lb; 606 lb
3. 80 times
4. 45 miles

5. 540–600 times
6. 1,130–1,170 times

OL Check student's word problem for accuracy and reasonableness.

BL 1,695–3,510

CHAPTER 5 Multiply with Two-Digit Numbers

Graphic Novel Basketball Brainteaser

168 points

OL Check student's word problem for accuracy and reasonableness.

BL Check student's graphic novel for accuracy and reasonableness.

Real-World Problem Solving Reader *Expanding the United States*

1. They traveled the same distance. Sample answer: 30 hours × 2 miles per maximum speed = 60 miles traveled for the covered wagon; 20 hours × 3 miles per hour = 60 miles traveled for walking.

2. 216 graves. Sample answer: 24 miles × 9 graves per mile = 216 graves.

3. $200 a week

4. 68 ounces of gold

5. $120

6. 900 houses; Thirty houses were made daily and there are approximately thirty days in a month.

Real-World Problem Solving Reader *What is Recycling?*

1. Sample answer: The chart shows data at different points in time. A timeline also shows data at different points in time. A timeline however is sequential and usually has equal intervals. This chart is sequential, but does not have equal intervals.

2. About 360 million tons

3. 175 trees; Sample answer: There are 25 students in our class, so I multiplied 25 by 7.

4. 51 trees

5. Sample answer: 540 pieces of paper; 36 weeks in a school year × 15 pieces of paper = 540 pieces of paper in a school year

BL Check student's work for accuracy.

 Activity Card Field Trip Fun

Check student's work for accuracy and reasonableness.

CHAPTER 6 Divide by a One-Digit Number

Activity Card Field Long Song Contest

Check student's work for accuracy.

Problem-Solving Card Desert Safari

1. about 140 lb

2. 300 miles

3. 640 lb

4. 2 ounces

5. hyena; 18 lb > 13 lb

6. 30 lb

7. giraffe: 60 lb, lion: 31 lb, hyena: 18 lb; giraffe, lion, hyena

OL Check student's word problem for accuracy and reasonableness.

BL

Animal	Amount of food per day (lb)	Amount per week
Hippopotamus	150	1,050
Elephant	160	1,120
Giraffe	60	420
Lion	About 217	About 1,519
Camel	About 90	About 630
Hyena	18	126
Chimpanzee	13	91
Flamingo	2 ounces	14 ounces

CHAPTER 7 Patterns, Sequences, and Functions

 Graphic Novel A Successful Season

Jamie 16; Carlos 6; Nate 16; Luke 10; Jacob 10; 58 points

OL Check student's word problem for accuracy and reasonableness.

BL Check student's graphic novel for accuracy and reasonableness.

 Activity Card Field Exercise Equations

Check student's word problems for accuracy.

 Problem-Solving Card A Visit to the Supermarket

1. 15 items

2. $3 + 5 + 2 = 10$

3. Add 4.

4. $17 - 2 = 15$

5. $12 \div 2 = 6$ bottles of juice

6. $4

Real-World Problem Solving Reader *Life in the United States*

1. $\frac{9}{50}$

2. $\frac{28}{50}$; There are 12 states in the Midwest region and 16 states in the South region. $12 + 16 = 28$

3. about $\frac{1}{10}$

4. $\frac{3}{4} > \frac{1}{2}$; I know that $\frac{3}{4} = 75\%$, and that $\frac{1}{2} = 50\%$.

5. about 14 times

6. about 270,000,000

CHAPTER 9 Fractions

 Activity Card Voting Age

Check student's work for accuracy.

 Activity Card I Will Take Two, Please

Banana Bread

4 eggs   1 tsp. salt

$3\frac{1}{2}$ c. sifted flour $\frac{2}{3}$ c. vegetable shortening

4 tsp. baking powder $1\frac{1}{3}$ c. sugar

$\frac{1}{2}$ tsp. baking soda   2 c. mashed bananas

Serves 32 people

BL

Banana Bread

8 eggs   2 tsp. salt

7 c. sifted flour $1\frac{1}{3}$ c. vegetable shortening

8 tsp. baking powder $2\frac{2}{3}$ c. sugar

1 tsp. baking soda   4 c. mashed bananas

Serves 48 people

CHAPTER 10 Fractions and Decimals

 Activity Card Olympic Race

Check to see if student's score cards are in order from greatest to least.

Problem-Solving Card Decimal Note-ation

1. 1.0

2. 0.5

3. 0.25

4.

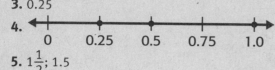

0   0.25   0.5   0.75   1.0

5. $1\frac{1}{2}$; 1.5

6. $1\frac{3}{4}$; 1.75

7. $4\frac{1}{4}$; 4.25

8. ♩

9. ♩

OL Check student's word problem for accuracy and reasonableness.

BL $\frac{1}{8}$, 0.125

CHAPTER 11 Customary Measurement

Real-World Problem Solving Reader *Strange but True*

1. 64 ounces ÷ 16 ounces per pound = 4 pounds; 64 ounces < 8 pounds

2. roadrunner: 15 miles per hour × 4 hours = 60 miles; cheetah: 70 miles per hour × 4 hours = 280 miles; 60 miles < 280 miles

3. 1 ounce < 9 pounds

4. 12 Kitti's hog-nosed bats; Sample answer: determine the wingspan of the golden-crowned flying fox bat in inches 6 feet × 12 inches per foot = 72 inches, determine the length of the Kitti bat 8 inches − 2 inches = 6 inches, 72 ÷ 6 = 12

5. 700 pounds

6. 8 minutes; Sample answer: 360 ants ÷ 45 ants per minute = 8 minutes

Real-World Problem Solving Reader *Ancient Giants of the Forest*

1. Florida Strangler Fig; 44 pounds

2. Florida Strangler Fig and Rocky Mountain Ponderosa Pine

3. Sample answer: Wrap a string around the base of the cylinder, and mark the string to show one time around. Measure the string.

4. 600 points

5. Sample answer: Measure the distance from the lowest branch to the ground. A tree earns 1 point for each inch. Each branch earns 2 points. Add the points together to find the total score. The equation would be: (1 × inches from lowest branch to ground) + (2 × number of branches) = total points

Problem-Solving Card Tide Pool Ecosystems

1. 80 times
2. 100 times
3. 40 times
4. length = 2 feet, width = 1 foot

5. length = 4 feet, width = 1 foot
6. 320 ounces
7. 400 ounces

OL 480 ounces, 640 ounces

BL Check student's word problem for accuracy and reasonableness.

CHAPTER 12 Metric Measurement

 Activity Card Gas Conservation

Check student's work for accuracy and reasonableness.

 Problem-Solving Card Coral Reefs

1. hard coral and soft coral
2. whale shark and dolphin
3. sea cucumber
4. jelly fish
5. whale shark
6. 2,000 millimeters
7. Check student's drawing.

OL Check student's work for accuracy and reasonableness

BL Check student's word problem for accuracy and reasonableness.

CHAPTER 13 Perimeter and Area

 Graphic Novel Having a Class Pet

30-gallon aquarium; Sample answer: The 20-gallon aquarium's floor area equals 360 square inches, which is too small. Both the 30-gallon and 40-gallon aquarium have a floor area that is at least 576 square inches, but the 30-gallon aquarium take up less room.

OL **BL** Check the group's graphic novel for accuracy and reasonableness.

Problem-Solving Card Walls with History

1. 4,000 square ft

2. 10,952 square ft

3. 1,500 square ft

4. about 900 yd

5. 2,240 square ft

6. 614 ft

7. Sample answer: Fort McIntosh; its dimensions are measured in yards, whereas Stone Fort at Harper's Ferry's dimensions are measured in feet.

OL 200 feet; 2,500 square feet

CHAPTER 14 Geometry

 Activity Card Symmetry in Nature

Check student's drawings for line of symmetry.

 Problem-Solving Card Symmetry in Nature

1. all of them

2. yes; 1

3. Sample answer: You can look to see if one half of the object is the mirror image of the other half. You can also decide if the image would look the same after it is turned.

4. Check student's drawings for line of symmetry.

5. Sample answer: Ripples or waves in the water.

OL Check student's word problem for accuracy and reasonableness.

BL Check student's work for accuracy and reasonableness.

Answers

Use the information to solve the problem.

Wet Weekend Fun!

Remember, I am going to the water park this weekend, and want to go to the most popular park.

Morgan, look at this! Slide Island's website says they had 89,868 visitors last year.

Well, my favorite, Wave City, had 79,416 visitors last year.

Which water park is Morgan going to visit this weekend? Explain.

_____

_____

_____

Morgan, Kendra, and Carlos in
Wet Weekend Fun!

Make a Big Difference
Subtract Multi-Digit Numbers

Name _____ Date _____

Use the information to solve the problem.

Dog Walking Dollars

Will Teresa earn enough money walking three dogs in two weeks to buy the DVD? Explain.

_____

_____

_____

Blackline Masters

Teresa and Ethan *in* Dog Walking Dollars

High and Low Game
Find a Product

```
___ ___ ___ ___          ___ ___ ___ ___          ___ ___ ___ ___
 X        ___              X        ___              X        ___
_____               _____               _____
```

```
___ ___ ___ ___          ___ ___ ___ ___          ___ ___ ___ ___
 X        ___              X        ___              X        ___
_____               _____               _____
```

```
___ ___ ___ ___          ___ ___ ___ ___          ___ ___ ___ ___
 X        ___              X        ___              X        ___
_____               _____               _____
```

```
___ ___ ___ ___          ___ ___ ___ ___          ___ ___ ___ ___
 X        ___              X        ___              X        ___
_____               _____               _____
```

Use the information to solve the problem.

How many minutes are still available on Julian's MP3 player? Explain.

_____

_____

_____

Julian and Olivia *in* Musical Math

Use the information to solve the problems.

A Pressing Problem

Remember, we are trying to figure out how many jelly beans the machine can press in 5 seconds.

The guidebook says that this machine can press 1,260 jellybeans in just one second!

1. How many jelly beans would Kendra like to go home with?

   _____

2. Suppose both Alyssa and Kendra went home with that many jelly beans. How many jelly beans would that be?

   _____

Alyssa & Kendra in
A Pressing Problem

Use the information to solve the problem.

How many points has Carmen scored during the 14-game basketball season?

_____

Name _____   Date _____

Carmen and Ashley ⁱⁿ *Basketball Brainteaser*

Name _____ Date _____

Use the information to solve the problems.

A Successful Season

Remember, Coach is sharing our individual stats for the season.

Player	Game 1	Game 2	Game 3	Game 4	Game 5
Jaime	1	2	4	8	
Carlos	6	6	6	6	
Nate	8	10	12	14	
Luke	10	10	10	10	
Jacob	14	13	12	11	
Total	35	37	40	45	

If the scoring pattern continues, how many points should each player on the Tigers team score at tomorrow's game? What will be the total for the game?

_____

The Gang in A Successful Season

Mixed Number Match Game, Sheet 1

$4\frac{3}{4}$	$1\frac{1}{3}$
$4\frac{1}{5}$	$1\frac{2}{3}$
$2\frac{2}{5}$	$4\frac{1}{4}$
$3\frac{3}{5}$	$2\frac{2}{4}$

Name _____ Date _____

Mixed Number Match Game, Sheet 2

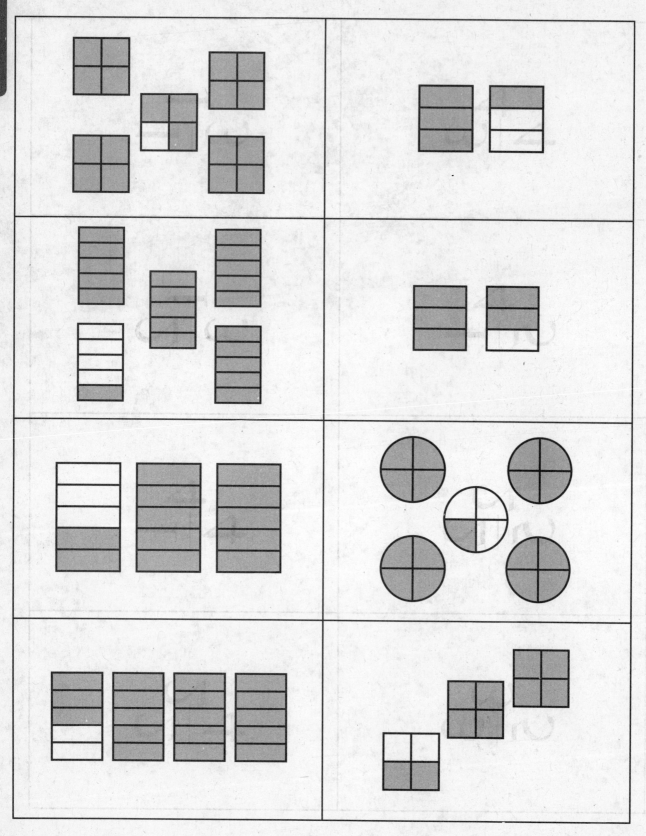

Mixed Number Match Game, Sheet 3

$3\frac{4}{5}$	$2\frac{4}{6}$
$3\frac{1}{6}$	$2\frac{5}{6}$
$1\frac{2}{6}$	$1\frac{1}{8}$
$2\frac{3}{6}$	$3\frac{2}{8}$

Mixed Number Match Game, Sheet 4

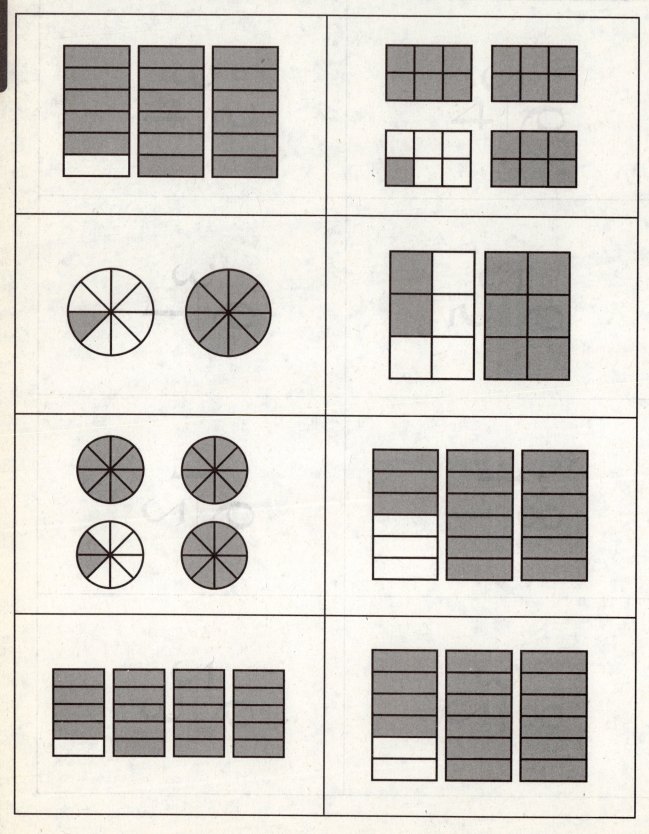

Mixed Number Match Game, Sheet 5

$3\frac{7}{8}$	$2\frac{3}{8}$
$3\frac{1}{9}$	$3\frac{4}{8}$
$1\frac{2}{9}$	$3\frac{5}{8}$
$2\frac{3}{9}$	$1\frac{6}{8}$

Name _____ Date _____

Mixed Number Match Game, Sheet 6

Name _____ Date _____

Mixed Number Match Game, Sheet 7

$1\frac{8}{9}$	$1\frac{4}{9}$
$2\frac{1}{10}$	$2\frac{5}{9}$
$1\frac{2}{10}$	$2\frac{6}{9}$
$3\frac{3}{10}$	$3\frac{7}{9}$

Mixed Number Match Game, Sheet 8

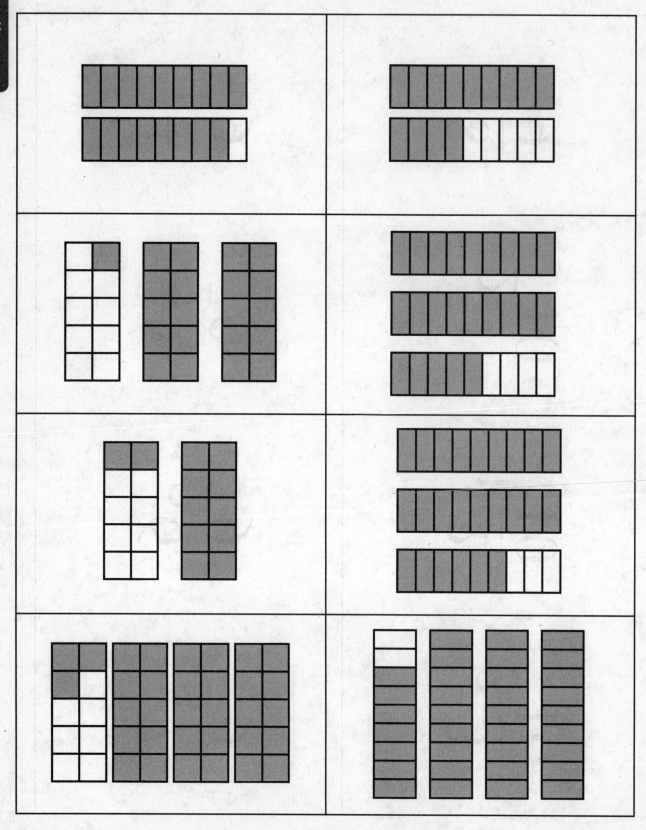

Mixed Number Match Game, Sheet 9

$1\dfrac{8}{10}$	$1\dfrac{4}{10}$
$1\dfrac{9}{10}$	$1\dfrac{5}{10}$
$3\dfrac{1}{2}$	$1\dfrac{6}{10}$
$5\dfrac{1}{2}$	$2\dfrac{7}{10}$

Mixed Number Match Game, Sheet 10

Fractoes and Decimoes Game, Sheet 1
Match Fractions and Decimals

$\frac{1}{10}$	**0.4**	$\frac{2}{10}$	**0.8**	$\frac{3}{10}$	**0.2**
eight-tenths	**0.1**	$\frac{10}{10}$	**0.3**	ten-tenths	**0.6**
0.5	$\frac{4}{10}$	**0.9**	$\frac{8}{10}$	**0.1**	$\frac{7}{10}$
0.8	$\frac{6}{10}$	**0.7**	$\frac{5}{10}$	**0.2**	$\frac{9}{10}$
0.3	six-tenths	**0.4**	one-tenth	**0.5**	two-tenths
0.6	seven-tenths	**0.7**	four-tenths	**0.8**	five-tenths
nine-tenths	**0.1**	three-tenths	**0.9**	one-tenths	$\frac{7}{10}$
two-tenths	$\frac{8}{10}$	three-tenths	**1.0**	four-tenths	**0.6**
[bar model]	**0.2**	five-tenths	[bar model]	$\frac{10}{10}$	[bar model]
six-tenths	[bar model]	[bar model]	seven-tenths	[bar model]	$\frac{9}{10}$

Name _____ Date _____

Fractoes and Decimoes Game, Sheet 2
Match Fractions and Decimals

0.5	(grid)	nine-tenths	(grid)	eight-tenths
0.4	(grid)	(grid)	**0.3**	ten-tenths
(grid)	**0.2**	six-tenths	(grid)	$\frac{4}{10}$
(grid)	five-tenths	(grid)	nine-tenths	$\frac{1}{10}$
0.3	seven-tenths	eight-tenths	$\frac{5}{10}$	$\frac{3}{10}$
$\frac{6}{10}$	two-tenths	**0.4**	ten-tenths	**1.0**	three-tenths

Area Guess
Find Area of Rectangles

Player _____

Object	Area		Difference
	Estimated	Actual	

Player _____

Object	Area		Difference
	Estimated	Actual	

Use the information to solve the problem.

Having a Class Pet

Remember, we want to get a guinea pig for a class pet.

So we need to choose a glass aquarium that is the right size.

I also found out that a guinea pig needs at least 576 square inches of living space.

Barney's Pet Shop
Aquariums

Size	Dimensions (length and width in inches)
20 gallons	30 x 12
30 gallons	36 x 18
40 gallons	48 x 12

Which aquarium has a floor area of at least 576 square inches and will take up the least amount of room in the classroom? Explain.

_____

_____

_____

Julian in Having a Class Pet